YOUR KNOWLEDGE HAS VALUE

Islam Ud Din Khan

Antimicrobial resistance of organisms causing neonatal sepsis

GRIN Verlag

Bibliografische Information der Deutschen Nationalbibliothek:

Die Deutsche Bibliothek verzeichnet diese Publikation in der Deutschen National-
bibliografie; detaillierte bibliografische Daten sind im Internet über http://dnb.d-
nb.de/ abrufbar.

Imprint:

Copyright © 2013 GRIN Verlag GmbH
Druck und Bindung: Books on Demand GmbH, Norderstedt Germany
ISBN: 978-3-656-56317-4

This book at GRIN:

http://www.grin.com/en/e-book/266185/antimicrobial-resistance-of-organisms-
causing-neonatal-sepsis

GRIN - Your knowledge has value

Der GRIN Verlag publiziert seit 1998 wissenschaftliche Arbeiten von Studenten, Hochschullehrern und anderen Akademikern als eBook und gedrucktes Buch. Die Verlagswebsite www.grin.com ist die ideale Plattform zur Veröffentlichung von Hausarbeiten, Abschlussarbeiten, wissenschaftlichen Aufsätzen, Dissertationen und Fachbüchern.

Visit us on the internet:

http://www.grin.com/

http://www.facebook.com/grincom

http://www.twitter.com/grin_com

Antimicrobial resistance of organisms causing neonatal sepsis

Abstract

Objective: To investigate the spectrum of organisms causing neonatal sepsis in Peshawar, Pakistan and to assess their sensitivity to various groups of drugs.

Methods: Blood taken from newborn babies admitted to the special care baby unit at the Khyber Teaching Hospital with a clinical diagnosis of neonatal sepsis was cultured. The data obtained from October 1997 to December 2000 were analysed and the results tabulated.

Results: A total of 1598 blood cultures were taken; 1003 were positive (positivity rate 62.8%).*Escherichia coli* was the most common organism found (36.6%), followed by *Staphylococcus aureus* (29.5%), *Pseudomonas* (22.4%), *Klebsiella* (7.6%), and *Proteus* (3.8%). No group B streptococcus was grown. *Listeria monocytogenes* was found in one cerebrospinal fluid culture. *E coli* and *Pseudomonas* showed a high degree of resistance to commonly used antibiotics (ampicillin, augmentin, and gentamicin), a moderate degree of resistance to cephalosporin (cefotaxime, ceftzidime, and ceftrioxone), and low resistance to drugs not used for newborn babies (ofloxacin, ciprofloxacin, and enoxabid). *S aureus* showed a low resistance to all three groups of antibiotics.

Conclusion: Neonatal sepsis remains one of the leading causes of neonatal admission, morbidity, and mortality in developing countries. Gram negative organisms are the major cause of neonatal sepsis in Peshawar. Such organisms have developed multidrug resistance, and management of patients infected with them is becoming a problem in developing countries.

Introduction

Neonatal sepsis is one of the most common reasons for admission to neonatal units in developing countries.[1,2] It is also a major cause of mortality in both developed and developing countries.[2-5] The spectrum of organisms that cause neonatal sepsis changes over time and varies from region to region. It can even vary from hospital to hospital in the same city. This is due to the changing pattern of antibiotic use and changes in lifestyle. Gram negative organisms were the most common cause of neonatal sepsis in Europe and America in the 1960s. It changed to group B streptococcus during the 1970s and coagulase negative *Staphylococcus* during the late 1980s and 1990s. In most of the developing countries, Gram negative organisms remain the major cause of neonatal sepsis, particularly early onset neonatal sepsis.[1,4-10] These organisms have developed increasing multidrug resistance over the last two decades,[8,9,11] due to the indiscriminate and inappropriate use of antibiotics, over the counter sale of antibiotics, lack of legislation to control their use, poor sanitation, and ineffective infection control in the maternity services.[12] The rapid emergence of multidrug resistant neonatal sepsis in developing countries is a new potential threat to the survival of newborn babies, who are often already in a poor condition.

MATERIALS AND METHODS

This was a prospective study carried out in the special care baby unit of the Khyber Teaching Hospital, Peshawar, which is a 1100 bed teaching hospital acting as a tertiary care centre for the rest of the province. More than 70% of admissions to the unit are outborn. The annual number of admissions is about 1700 per year. Sepsis accounts for almost 40% of the admissions. Babies with a clinical diagnosis of neonatal sepsis were enrolled in this study. Babies who had received antibiotics before admission or had surgical problems, chromosomal and congenital anomalies, or dysmorphism were excluded.

Blood was taken by a standard method and cultured in the microbiology laboratories of Pakistan Medical Research Council, Khyber Medical College, Peshawar. Sensitivity to various antibiotics was tested by a standard disc diffusion technique.

RESULTS

Blood culture results obtained from October 1997 to December 2000 were analysed. Of a total of 1598 blood cultures, 1003 were positive (positivity rate of 62.8%): 367 (36.6%) were positive for *Escherichia coli*, 296 (29.5%) for *Staphylococcus aureus*, 225 (22.4%) for *Pseudomonas*, 77 (7.6%) for *Klebsiella*, and 38 (3.8%) for *Proteus*. No group B streptococcus was grown from any culture, and *Listeria monocytogenes* was grown from one cerebrospinal fluid culture.

The pattern of sensitivity of these organisms was analysed for three groups of antibiotics:

1. penicillins and aminoglycosides, which are used as first line antibiotics;

2. cephalosporins, which are used as a second line antibiotics;

3. quinolones, which are not recommended for use in children less than 4 years of age, but may be indicated if the child has blood culture positive severe sepsis and the organism is not sensitive to any other antibiotic.

Table 1 shows the pattern of sensitivity of *E coli*, *S aureus*, *Pseudomonas*, *Klebsiella* sp, and*Proteus* to various antibiotics. Considerable resistance to first line antibiotics, moderate resistance to cephalosporins, and low resistance to quinolones was observed.

Table 1: Pattern of resistance to various antibiotics

Antibiotic		*E coli*	*S aureus*	*Pseudomonas*sp	*Klebsiella*sp	*Proteus*sp
Values are percentages.						
S, Sensitive; R, resistant.						
First line						
Ampicillin	S	11	60	13.57	34.66	47.5
	R	89	40	86.43	65.34	52.5
Gentamicin	S	21.3	30	21.4	16	18
	R	78.7	70	78.6	84	82
Augmentin	S	6.5	27	4.48	18.75	39.6
	R	93.5	73	95.52	81.25	60.4
Second line						
Cefotaxime	S	32.6	50	27	14	31.5
	R	67.4	50	73	86	68.5
Ceftazidime	S	32.5	36.76	43.49	33.76	28.9
	R	67.5	63.24	56.51	66.24	71.1
Ceftrioxone	S	28	41.23	33	18	34.28
	R	72	58.77	67	82	65.72
Quinolones						
Ofloxacin	S	76.43	73.35	73	–	84
	R	23.57	26.65	27		16
Ciprofloxacin	S	57.5	59	77	–	55.55
	R	42.5	41	23		44.45
Enoxabid	S	60	44	49.6	–	–
	R	40	56	50.4		

DISCUSSION

About five million neonatal deaths occur worldwide every year, 98% of which occur in developing countries, particularly Asia and Africa. Infections such as tetanus, pneumonia, septicaemia, meningitis, and diarrhoea account for 30–50% of neonatal deaths in developing countries.[13] Neonatal sepsis is a life threatening emergency and any delay in treatment may result in death.[14]

The spectrum of organisms causing neonatal sepsis in our study is similar to that reported for other neonatal units in developing countries, with Gram negative organisms being responsible for most cases, particularly early onset. Almost 70% of episodes of neonatal sepsis in our unit are caused by Gram negative organisms, with *E coli* being the most common (36.6%) and *Pseudomonas* the second most common (22.4%). A similar pattern has been reported for the Children's Hospital, Lahore.[9] In that series, Gram negative organisms were responsible for almost 80% of episodes of neonatal sepsis, with *E coli* being the most common (45.8%) followed by *Klebsiella* (17.2%) and *Pseudomonas* (16.2%). Bhutta and Yusuf[4] reported that *Klebsiella* was the most common cause of neonatal sepsis in Karachi, Pakistan. Joshi *et al*,[8] from India, reported Gram negative sepsis in 67.2% of their cases, with *Pseudomonas aeruginosa* being the most common organism (38.3%) followed by *Klebsiella* (30.4%) and *E coli* (15.6%). Similar patterns have been reported in Trinidad[5] and Southern Israel.[6]

S aureus was the second most common organism in our study. Anwer *et al*[1] found Gram positive organisms to be the main cause of neonatal sepsis in a teaching hospital in Karachi, Pakistan. Gram negative organisms were responsible for almost half of the episodes of early onset neonatal sepsis in their series. Similar results have been reported by Dawodu *et al*[2] and Kilani and Basamad[10] for Riyadh, Saudi Arabia.

Group B streptococcus was not isolated from any culture in our series. The same has been reported in most of the studies from Pakistan and other developing countries.[1,2,4,8,9] Ghiorgis,[15] in his study from Ethiopia, did not find any group B streptococcus. On the other hand, in the series reported by Robbilard *et al*[16] from Guadeloupe, group B streptococcus was grown from 46% of positive blood cultures, and 52% of gastric aspirates were positive for group B streptococcus. In Al Wasl Hospital, Dubai, Koutouby and Habib Ullah[17] found 106 culture positive cases of neonatal sepsis, with group B streptococcus being the most common organism (23%) particularly in early onset and very early onset neonatal sepsis. Ohlsson *et al*[18] were the first to report the emergence of group B streptococcus in Saudi Arabia in the early 1980s.

Multidrug resistance of the causative organisms of sepsis is a rapidly emerging, potentially disastrous problem.[19,20] Our data show that Pakistan is no exception to this worldwide antimicrobial emergency. In fact, the situation is worst in developing countries because of the lack of control of the use of antibiotics, the non-existence of legislation on antibiotic prescription, over the counter sale of antibiotics, poor sanitary conditions, lack of basic facilities and practices such as hand washing, lack of surveillance of the standards of maternity homes, and the practices of traditional birth attendants, who deliver almost 80% of all babies.[21]

Our study shows a very high degree of resistance of Gram negative organisms to first line antibiotics. About 40% of *S aureus* were resistant to ampicillin. There is also high degree of resistance to cephalosporins by both Gram positive and Gram negative organisms. Only 43.5% of *Pseudomonas* were sensitive to ceftazidime. There is low degree of resistance to quinolones, particularly ofloxacin. This is a relatively new class of antibiotics, the use of which has recently become very common, particularly in general practice. Similar results have been reported from other parts of Pakistan. The pattern of sensitivity reported by Maryam et al[9] for the Children's Hospital in Lahore is similar to ours except that *S aureus* and *Staphylococcus epidermidis* were found to be much more resistant to quinolones.

The data of Anwer et al[1] from Karachi show 80% resistance to ampicillin but only 11–13% resistance to cefotaxime and 0–10% resistance to amikacin. Bhutta et al[22] from Karachi also reported a high degree of resistance to ampicillin and gentamicin among Gram negative organisms.

Emerging multiple drug resistance has also been reported in other parts of the world. The data of Orrett and Shurland[5] from Trinidad show 85% of *S aureus* are resistant to ampicillin, and *Pseudomonas* had 76.6% resistance to ceftazidime and 72.1% resistance to gentamicin. The study of Joshi et al[8] from India shows a predominance of Gram negative bacteraemia (67.2%) in their series, which had 25–75% resistance to cephalosporins, 68–78% resistance to piperacillin, and 23–69% resistance to gentamicin.

Friedman et al[11] from Toronto isolated ampicillin resistant *E coli* from 75% of infants with early onset neonatal sepsis and 53% from a group with late onset neonatal sepsis. Gentamicin resistance was found in 50% of the early onset group and 16% of the late onset group. Kaushik et al[23] reported their bacterial isolates to be resistant to penicillin, ampicillin, and gentamicin, but with good sensitivity to third generation cephalosporins and netilmicin. Leibovitz et al[24] reported the appearance of extremely virulent, multiresistant *Klebsiella* in their neonatal intensive care unit in Kaplan Hospital, Israel. Koksal et al,[25] from India, reported a series of 35 cases of severe Gram negative neonatal sepsis, with all the organisms being resistant to ampicillin, amoxicillin, ticarcillin, cefazoline, cefotaxime, ceftazidime, ceftrioxone, and aminoglycoside. They treated these babies with meropenem and achieved 94.3% satisfactory clinical and bacterial response. The routine use of intrapartum antibiotic prophylaxis for the prevention of group B streptococcus septicaemia in newborn babies has resulted in the appearance of ampicillin resistant Gram negative neonatal sepsis in a large number of developed countries.[26–28]

Antibiotic resistance is increasing world wide and has become a serious health problem in hospitals and the community. Infection with resistant organisms has been associated with treatment failure, higher morbidity and mortality, and increased costs. This has necessitated the development, implementation, and evaluation of policies on the use of antibiotics.[13,19,29] Prudent use of antibiotics and antibacterials must be promoted to maintain the balanced microbial environment in which we live.[30] Routine bacterial surveillance and study of their resistance patterns must be an essential component of neonatal care. A knowledge of these patterns is essential when local policies on the use of anitbiotics are being devised.

REFERENCES

1. **Anwer SK**, Mustafa S, Pariyani S, *et al*. Neonatal sepsis: an etiological study. *J Pak Med Assoc*2000;50:91–4.

2. **Dawodu A**, Al-Umran K, Twum-Danso K. A case control study of neonatal sepsis: experience from Saudi Arabia. *J Trop Pediatr*1997;43:84–8.

3. **Stoll BJ**, Holman RC, Schuchat A. Decline in sepsis associated neonatal and infant deaths in the United States, 1979 through 1994. *Pediatrics*1998;102:e18.

4. **Bhutta ZA**, Yusuf K. Neonatal sepsis in Karachi: factors determining outcome and mortality. *J Trop Pediatr*1997;43:65–70.

5. **Orrett FA**, Shurland SM. Neonatal sepsis and mortality in a regional hospital in Trinidad: aetiology and risk factors. *Ann Trop Paediatr*2001;21:20–5.

6. **Greenberg D**, Shinwell ES, Yagupsky P, *et al*. A prospective study of neonatal sepsis and meningitis in southern Israel. *Pediatr Infect Dis J*1997;16:768–73.

7. **Ghiorghis B**. Neonatal sepsis in Adis Ababa, Ethiopia: a review of 151 bacteremic neonates. *Ethiop Med J*1997;35:169–76.

8. **Joshi SJ**, Ghole VS, Niphadkar KB. Neonatal gram negative bacteremia. *Indian J Pediatr*2000;67:27–32.

9. **Maryam W**, Laeeq A, Maqbool S. Neonatal sepsis spectrum of antibiotic resistance.*Proceedings of 10th Annual National Pediatric Conference*. 2001;57.

10. **Kilani RA**, Basamad M. Pattern of proven bacterial sepsis in a neonatal intensive care unit in Riyadh-Saudia Arabia: 2 year analysis. *J Med Liban*2000;48:77–83.

11. **Friedman S**, Shah V, Ohlsson A, *et al*. Neonatal Escherichia coli infections: concerns regarding resistance to current therapy. *Acta Paediatr*2000;89:686–9.

12. **Rahman S**, Roghani MT, Ullah R, *et al*. A survey of perinatal care facilities in Pakistan.*Proceedings of 10th National Annual Pediatric Conference* 2001;34.

13. **Darmstadt GL**. Global newborn health challenges and opportunities. *Proceedings of 10th National Annual Pediatric Conference* 2001;22.

14. **Yurdakok M**. Antibiotic use in neonatal sepsis. *Turk J Pediatr*1998;40:17–33.

15. **Ghiorgis B**. Neonatal sepsis. *Ethiop Med J*1991;28:956–7.

16. **Robbilard PY**, Perez JM, Hulsey TC, *et al*. Evaluation of neonatal sepsis screening in a tropical area. Part 1: major risk factors for bacterial carriage at birth in Guadeloupe. *West Indian Med J*2000;49:312–15.

17. **Koutouby A**, Habib Ullah J. Neonatal sepsis in Dubai United Arab Emirates. *J Trop Pediatr*1995;41:177–80.

18. **Ohlsson A**, Bailey T, Takieddine F. Changing etiology and outcome of neonatal septicemia in Riyadh, Saudi Arabia. *Acta Paediatr Scand*1986;75:540–4.

19. **Pennington H**. Millenium bugs. *Biologist (London)*2000;47:93–5.

20. **Bax R**, Mullan N, Verhoef J. The millenium bugs: the need for and development of new antibacterials. *Int J Antimicrob Agents*2000;16:51–9.

21. **Faryal FF**, Rahbar HM, Ali T. Traditional newborn practices in selected low socioeconomic settlements of Karachi, Pakistan. *Proceedings of 10th National Annual Pediatric Conference*. 2001;23.

22. **Bhutta ZA**, Naqvi SH, Muzaffar T, *et al*. Neonatal sepsis in Pakistan. Presentation and pathogens. *Acta Paediatr Scand*1991;80:596–601.

23. **Kaushik SL**, Parmer VR, Grover N, *et al*. Neonatal sepsis in hospital born babies. *J Commun Dis*1998;30:147–52.

24. **Leibovitz E**, Flidel-Rimon O, Juster Reicher A, *et al*. Sepsis in a neonatal intensive care unit: a four year retrospective study (1989–1992). *Isr J Med Sci*1997;33:734–8.

25. **Kokasal N**, Hacimustafaoglu M, Bagei S, *et al*. Meropenem in severe infections due to multiresistant gram negative bacteria. *Indian J Pediatr*2001;68:15–19.

26. **Terrone DA**, Rinehart BK, Einstein MH, *et al*. Neonatal sepsis and death caused by resistant Escherichia coli: possible consequences of extended maternal ampicillin administration. *Am J Obstet Gynecol*1999;180:1345–8.

27. **Levin EM**, Ghai V, Barton JJ, *et al*. Intrapartum antibiotic prophylaxis increases the incidence of gram negative neonatal sepsis. *Infect Dis Obstet Gynecol*1999;7:210–13.

28. **Mercer BM**, Carr TL, Beazley DD, *et al*. Antibiotic use in pregnancy and drug resistant infant sepsis. *Am J Obstet Gynecol*1999;181:816–21.

29. **Goossen H**. Antibiotic resistance and policy in Belgium. *Verh K Acad Geneeskd Belg*2000;62:439–69.

30. **Levy SB**. Antibiotic and antiseptic resistance: impact on public health. *Pediatr Infect Dis*2000;19(suppl 10):S120–2.